I0469300

Preface

This book is written as a supplement to Process Design for Chemical Engineers with following additions for each of the eight chapters: (1) comments or additional information are provided, (2) exercises are provided, which can be used for readers to test their understanding or for professor to assign them to students as homework, and (3) examples are provided to illustrate some design technique and calculation.

Any comment to this book, please send an e-mail to this address: yu.processdesign@gmail.com.

Frank Chi-Liang Yu, author, Ten Books, Inc. December 25, 2012

Disclaimer

It is user's responsibility to use any information in this book. Every effort has been made by the author to assure the accuracy and reliability of the data contained in this book; however, the author and the book's owner make no representation, warranty, or guarantee in connection with this publication and hereby expressly disclaim any liability or responsibility for loss or damage resulting from its use.

Notes:
1. This book's ISBN numbers are: ISBN-13: 978-1481928328, ISBN-10:1481928325.
2. This book has copyright protection.
3. This book is owned and published by Ten Books, Inc., with CreateSpace as a distributor.

Revision history of Process Design for Chemical Engineers Supplement:
1st revision: October, 2014

Table of content

Chapter 1 Process Design

I. Comment: After study this chapter, reader should know what tasks will be done at different phases of an engineering project.

II. Exercise

1. Most engineering projects can be classified into two types. What are they?
2. Name the six phases of an engineering project.
3. Is it true that every engineering project has six phases?
4. Name the phases of an engineering project which a process design engineer may be involved.
5. Name the typical process design engineer's work in phase 1, 2, and 3?
6. What is process flow diagram (PFD)?
7. What is material balance? What is its purpose? What is the major information in it? How is it generated?
8. What is material selection diagram? What is its purpose? How is it generated?
9. What is equipment data sheet? What is its purpose? How is it generated?
10. What is instrument data sheet? What is its purpose? How is it generated?
11. What is pressure relief study (contingency analysis)? What is its purpose?
12. What is pressure relief device data sheet? What is its purpose? How is it generated?
13. What is HAZOP analysis? What is its purpose?
14. What is line list? What is its purpose? How is it generated?
15. Name two kinds of economic study in an engineering project.

Chapter 2 Pump

I. Comment

I.1 After study this chapter, reader should understand the basic features of centrifugal pump, reciprocating pump, and rotary pump, and be able to 1) calculate the basic pump performance parameters, such as required pump head, pump NPSHA, and required pump horsepower, 2) select a new pump among centrifugal, reciprocating, or rotary pump, 3) rate an existing centrifugal or reciprocating pump, and 4) operate a pump properly.

I.2 NPSH margin: Interested reader should contact Hydraulic Institute for the latest development. Reference 1 mentioned that for most pump applications, NPSH margin is at least ten feet.

I.3 Make sure to operate the centrifugal pump close to or within acceptable range of BEP to avoid recirculation in pump.

II. Exercise

1. What is the purpose of using a pump?
2. Name at least 5 process information should be provided on a pump data sheet.
3. Name two major types of pump, their principle of pumping, and the reasons they are selected.
4. Name major components of a centrifugal pump.
5. Name 3 types of centrifugal pumps due to different impeller design. What is their specific speed range?
6. What information is provided in a centrifugal pump's pump curve?
7. How to find a pump's operating point?
8. What are NPSHA and NPSHR? Why is cavitation happened in some pumps?
9. Why some centrifugal pump will have recirculation problem? How to prevent it?
10. Why some centrifugal pump is sealless?
11. Name two major types of a positive displacement pump. What is the difference between them?
12. Name three types of reciprocating pump.
13. Name at least three types of rotary pumps.
14. How to make reciprocating pump discharge flow smoother instead of pulsation?
15. How to calculate required pump head?
16. Which equation should be used to estimate pumping horsepower requirement?
17. NPSHA should be calculated for each pump. Which equation should be used?
18. For centrifugal pump, can we use the same pump curve for different pumping fluids?
19. What is suction specific speed for a centrifugal pump? Why do we need this information?
20. What is pump affinity law for a centrifugal pump?
21. How to reduce the acceleration head required for reciprocating pump, if it has long suction line (>15 ft)?
22. For a centrifugal pump application, what is its pumping liquid's upper viscosity limit?

III. Example

Example 1 Pump head calculation (section IV.1): From line hydraulic analysis, it is found that pump suction pressure (Ps) is 10 psig, and pump discharge pressure (Pd) is 110 psig. Pumping fluid density is 40 lb/ft3. What is the required pump head (H) in feet?

DP = 110 -10 =100 psi - [calculate pump differential pressure (DP) in psi, using Eq. (1a) .]
H = 100 / (40 / 144) = 360 ft.   [1 feet liquid head = 40/144 psi]
Alt. calc.: liquid s.g. = 40/62.37 = 0.6413; H = 100 * 2.31 / 0.6413 = 360 ft. [Eq. (1b)]

Using the required pump head and pumping rate, a pump can be selected from a manufacturer's catalog.

Example 2 Pump type selection: What type of pump should be used for following service? The design pumping rate is 550 gpm at 100°F. Its density is 40 lb/ft3 and viscosity is 0.5 cp. (1) Required pump head is 300 feet, (2) Required pump head is 2000 feet, (3) pumping rate is 10 gpm, and required pump head is 1500 feet.

(1) From section III.1, a single stage centrifugal pump can be used for this service, since its required pump head (300 feet) is less than 1000 feet, and its design pumping rate (550 gpm) is less than 500,000 gpm. Therefore, we can search centrifugal pump manufacturer's catalog to find a pump suitable for this service.

(2) Since required pump head is greater than 1000 feet, a single stage centrifugal pump may not be able to handle the service. The next thing we need to check is pumping fluid's viscosity. Since liquid viscosity is below 110 centistokes, a larger impeller, a high speed, or a multistage centrifugal pump is required. Consult a pump engineer or manufacturer for selection. [Note: For liquid viscosity greater than 110 centistokes, either a reciprocating or rotary pump should be used.]

(3) For small flow and high head application, diaphragm pump is usually selected, see section III.2.1.1.3.

Consult a pump specialist or pump manufacturer, if not sure about how to select a pump for an application.

Example 3 NPSHA (net positive suction head available) calculation: A pump (G-501A/B) is used to pump liquid from a surge drum (D-501), operating at 5 psig, 100°F, to a tower and another surge drum. Pumping rate is 350 gpm. From suction line hydraulic study, suction side line loss is 0.5 psi. Pumping liquid's vapor pressure at 100°F is 10 psia, and its density is 40 lb/ft3. What is the NPSHA of this pump? For the elevation of D-501 and G-501A/B, see Figure 9.

(1) For a centrifugal pump, pump NPSHA is calculated using Eq. (4) as follows (section IV.3):
NPSHA = ((5+14.7) – 10 – 0.5) * 2.31 / 0.6413 + (10 – 2) = 41.1 feet
[Note: liquid level at vessel bottom is used to calculate NPSHA for conservative reason.]

(2) For a reciprocating pump, pump NPSHA is calculated as follows (section IV.7.2): [This section is used to explain how to calculate NPSHA of a reciprocating pump, assuming that G-501A/B is a reciprocating pump.]

Acceleration head (Ha) needs to be calculated first as follows, using Eq. (15):

Following information needs to be collected in order to calculate Ha:

1) suction line length (L) = 25 feet, 2) average suction line velocity (v) = 4 ft/sec, 3) pump speed (N) = 360 rpm, 4) for a triplex pump, C = 0.066, 5) for hydrocarbon liquid, k = 2.

Ha = 25 * 4 * 360 * 0.066 / (32.2 * 2) = 36.9 feet

Once Ha is calculated, NPSHA can be calculated using Eq. (16) as follows:

NPSHA = ((5+14.7) – 10 – 0.5) * 2.31 / 0.6413- Ha + (10 – 2) = 4.2 feet

This NPSHA can be increased by installing a pulsation damper at suction line, since Ha can be reduced to 15 ft.

When selecting a pump from pump catalog, make sure pump's NPSHR is less than NPSHA.

Example 4 Pump horsepower (BHP) calculation (section IV.2): The design pumping rate at 100°F is 550 gpm. Required pump head is 300 feet. Its density is 40 lb/ft3. (1) What is the required pump horsepower at design condition? (2) This pump's normal pumping rate is 500 gpm. What is the required pump horsepower at normal condition?

(1) Required pump horsepower at design condition: Assume a pump is selected from a pump manufacturer's catalog with pump curve like Figure 16. From Figure 16, this pump size is 3x4-10, which means pump suction nozzle is 3", its discharge nozzle is 4", and maximum pump impeller is 10". For the required pump head (300 feet) and design pumping rate (550 gpm), 9" diameter pump impeller will be adequate for the service. For 550 pumping rate, its pump head is 312 feet, and pump efficiency (E) is 0.714 (71.4%).

DP = 312 * (40/144) = 86.7 psi
BHP = 550 * 86.7 / (1714 * 0.714) = 38.9 hp   [Eq. (2)]

This BHP is the required pump horsepower at design condition.

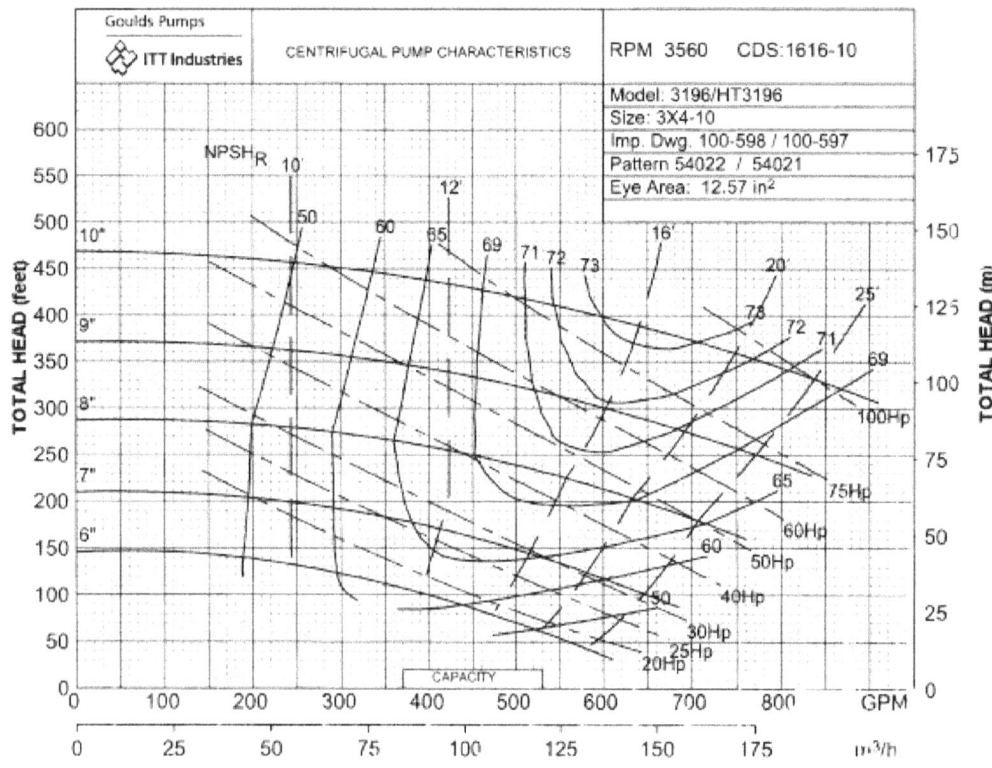

Figure 16 Pump curve for a selected centrifugal pump, for section III, Example 4. [Courtesy of ITT Goulds pumps, Seneca Falls, NY]

If we use 9" impeller, pump head generated at 550 gpm is 312 feet, greater than the required 300 feet. An alternate solution is to trim the pump impeller, so that pump head generated at 550 gpm is 300 feet. Using pump affinity law, Eq. (11e), the required new pump impeller diameter is calculated as follows.

D2 = (300/312)^(0.5) * 9 =8.825 inch – suggest to trim pump impeller diameter to 8.875 inch.
H = (8.875/9)^2 * 312 = 303 ft.
DP = 303 * (40/144) = 84.2 psi
BHP = 550 * 84.2 / (1714 * 0.714) = 37.8 hp – assume pump efficiency is still at 0.714.

Comparing to 9" impeller case, using 8.875" impeller will save 1.1 hp. Assume electricity cost is $0.1/kwh, and one year has 8000 hours. Annual saving is $656. If trimming this pump impeller costs more than $1968, it is not worth to trimming it (based on three year payout.).

(2) Required pump horsepower at normal operating condition: Assuming pump impeller size is 9". At normal pumping rate (500 gpm), pump head is 322 feet, and pump efficiency (E) is 0.704 (70.4%). Required pump horsepower can be calculated similar to (1).

DP = 322 * (40/144) = 89.4 psi
BHP = 500 * 89.4 / (1714 * 0.704) = 37.0 hp

This BHP is the required pump horsepower at normal operation. This is the normal electricity consumption of this pump.

Usually, pump data sheets will be sent to several pump vendors for them to provide a suitable pump and quote their pump price. After receiving all pump vendors' quotes, a bid evaluation table will be prepared by a pump engineer, and a recommendation of which pump to buy will be provided.

Example 5 Suction specific speed (S) calculation of a centrifugal pump (section IV.6.3): For the same pump application described in Example 3, S is calculated using Eq. (10). Assuming selected pump speed is 3550 rpm, and NPSHR is 20 feet.

S = 3550 * 350^0.5 / (20^0.75) = 7022 < 8500 – This pump S is acceptable per section III.1.7.

Example 6 Preliminary rating (check process capacity) of an existing pump for new service: From hydraulic study, we will know the required pump head and NPSHA for the new service.

(1) For centrifugal pump: From pump curve, we can check whether the existing pump is able to provide the required pump head or not, and whether the NPSHA is greater than the required NPSHR or not. If they are, the existing pump is able to handle the new service. The next thing to check is the existing pump driver's horsepower. Calculate the required pump horsepower as Example 4. If the required horsepower is less than the driver's design horsepower, the existing pump driver is adequate for the new service. Otherwise, a new larger driver will be required.

(2) For reciprocating pump: Using equations in section IV.7.1 to check existing pump's capacity for new service. Eq. (16) is used to calculate pump's NPSHA. Compare it with its NPSHR (if available) to check whether it is adequate for the new service. Calculate required pump horsepower per Eq. (2), if pump's efficiency is available for the new service. If it is less than the driver's horsepower, this pump will be adequate for the new service. If NPSHR and pump efficiency at new service are not available, consult pump vendor for these information.

(3) For rotary pump: Consult pump manufacturer or pump engineer to check the adequacy of an existing pump for new service.

Chapter 3 Compressor

I. Comment

I.1 After study this chapter, reader should understand the basic features of centrifugal compressor, reciprocating compressor, and rotary compressor, and be able to 1) calculate the basic compressor performance parameters, such as required compressor head, compressor discharge temperature, and required compressor horsepower, 2) select a new compressor among centrifugal, reciprocating, or rotary compressor, 3) preliminary rating of an existing compressor, and 4) operate a compressor properly.

I.2 For a centrifugal compressor, the lower limit of gas molecular weight may be as low as 5, depending on the manufacturer. In one application, for a gasoline hydrotreater, hydrogen rich recycle gas's molecular weight is about 5. A centrifugal compressor was selected for the service, due to its lower cost than an OIS (oil-injected screw compressor), and owner's preference.

II. Exercise

1. What is the purpose of using a compressor?
2. Name at least 5 process information should be provided on a compressor data sheet.
3. Name three major types of compressor, their principle of compressing, and the reasons they are selected.
4. Name two major types of compressor based on application.
5. Name major components of a single-stage and multistage centrifugal compressor.
6. Show the general shape of compressor performance curve for a centrifugal, axial, and reciprocating or rotary compressor?
7. Why a centrifugal compressor will surge? How to prevent it?
8. Name major components of an axial compressor.
9. Name two major types of a positive displacement compressor. What is the difference between them?
10. Name major components of a reciprocating compressor.
11. Name two types of reciprocating compressor by the compression element used.
12. Name major components of a rotary compressor.
13. Name at least three types of rotary compressor.
14. How to make reciprocating compressor discharge flow smoothly instead of pulsation?
15. For a centrifugal compressor application, what is the lowest limit of gas molecular weight?

III. Example

Example 1 A compressor is used to compress gas from 50 psig, 100°F to 150 psig. Design gas flow rate is 50 mmscfd (at 14.7 psia, 60°F), mw is 20. At inlet conditions, gas isentropic exponent (k) is 1.2, and its compressibility factor (Z) is 1.0. What type of compressor should be used? Calculate the required horsepower?

Assume compressor isentropic efficiency is 75% (0.75).

(1) Calculate gas flow at inlet condition:

Q (in acfm) = (50 x 10^6/(24*60)) (14.7/(50+14.7)) ((100+460)/(60+460)) = 8495.8 acfm

Alt. W (lb/min) = 50,000,000*20/(24*60)/379.6=1829.4
ρ1(gas density at suction, lb/ft3)=P1*MW/(Z1*R*T1)=(50+14.7)(20)/(1*10.731*(100+460))=0.2153
Q (in acfm) = 1829.4/0.2153 = 8495.8 acfm

(2) Type of compressor: From Fig. 2b, for this inlet flow rate and discharge pressure, a centrifugal

compressor, an oil-free rotary screw compressor, or an oil-injected rotary screw compressor can be used.

(3) Required horsepower:

Ps (suction pressure, psia) = 50+14.7=64.7; Pd (discharge pressure, psia) =150+14.7=164.7
r (compression ratio) = 164.7/64.7=2.5456 [Eq. (2b)]

Calculate discharge temperature using Eq. (4).

Td (discharge temperature, °R) = (100+460)+(100+460)(2.5456^((1.2-1)/1.2)-1)/0.75=685.8°R=225.8°F

Assume at this discharge conditions, gas isentropic exponent (k) is still at 1.2, and its compressibility factor (Z) is still at 1.0. Then, Td calculation is all right. Otherwise, an average isentropic exponent should be used to recalculate Td. Repeat the same calculation until calculated Td won't change anymore.

Let kx = (k-1)/k = (1.2-1)/1.2=0.1667

Hs (required isentropic head, feet) =1545(1)(100+460)(2.5456^0.1667-1)/(20*0.1667)=43737.8 [Eq. (2a)]

Assume a centrifugal compressor is selected. From Table 1, the compressor can be a single stage using an open impeller, or a multistage unit using several covered impellers (Fig. 8 shows that it needs 4 impellers).

GHP = 1829.4*43737.8/(33000*0.75) = 3232.9 hp [Eq. (4)]
BHP = GHP + GHP^0.4 = 3232.9+25.4=3258.3 hp [Eq. (10-11)]

Note: Instead of hand calculation, a computer simulator can be used to calculate the required compressor head, discharge temperature, gas physical properties (such as isentropic exponent, compressibility factor), and horsepower for a given compressor efficiency.

Example 2 A compressor is used to compress gas from 50 psig, 100°F to 500 psig. Design gas flow rate is 5 mmscfd (at 14.7 psia, 60°F), mw at 4. At inlet conditions, gas isentropic exponent (k) is 1.4, and its compressibility factor (Z) is 1.0. What type of compressor should be used? Calculate the required horsepower?

Assume compressor isentropic efficiency is 75% (0.75).

(1) Calculate gas flow at inlet condition:

Q (in acfm) = (5 x 10^6/(24x60)) (14.7/(50+14.7)) ((100+460)/(60+460)) = 849.6 acfm

Alt. W (lb/min) = 5,000,000*4/(24*60)/379.6=36.6
$\rho 1$(gas density at suction, lb/ft3)=P1*MW/(Z1*R*T1)=(50+14.7)(4)/(1*10.731*(100+460))=0.0431
Q (in acfm) = 36.6/0.0431 = 849.6 acfm

(2) Type of compressor: From Fig. 2b, for this inlet flow rate and discharge pressure, either a centrifugal compressor, a reciprocating compressor, or an oil-injected rotary screw compressor can be used. However, because of the low gas molecular weight, centrifugal compressor may not be a good choice. In this example, a reciprocating compressor will be used.

(3) Required horsepower:

Ps (suction pressure, psia) = 50+14.7=64.7; Pd (discharge pressure, psia) =500+14.7=514.7
r_t (overall compression ratio) = 514.7/64.7=7.955 [Eq. (2b)]

Since this compression ratio is over 5, two or three stage compression is required. Let's try two-stage compression first. Allow 5 psi for inter-stage cooler and piping line loss. For equal compression ratio at each stage, after trial and error, it is found that compression ratio for each stage (r_1 and r_2) is 2.86. Once compression ratio for each stage is known, the performance of each stage can be calculated as follows:

For first stage compressor:

Pd_1 (1st stage discharge pressure, psia) $= 2.86*(50+14.7) = 185.0$ [Eq. (2b)]

Calculate discharge temperature using Eq. (4).

Td_1 (1st stage discharge temperature, °R) $= (100+460)+(100+460)(2.86^{((1.4-1)/1.4)}-1)/0.75 = 821.5°R = 361.5°F$

Assume at this discharge conditions, gas isentropic exponent (k) is 1.3, and its compressibility factor (Z) is 0.9. The average isentropic exponent and compressibility factor for the 1st stage compression are as follows:

$k1$ (avg) $= (1.4+1.3)/2 = 1.35$; $Z1$ (avg) $= (1+0.9)/2 = 0.95$;   Let $kx1 = (k1-1)/k1 = (1.35-1)/1.35 = 0.2593$

Recalculate Td_1(1st stage discharge temperature, °R):

$Td_1 = (100+460)+(100+460)(2.86^{((1.35-1)/1.35)}-1)/0.75 = 793.8°R = 333.8°F$

Assume at this discharge conditions, gas isentropic exponent (k) is still 1.3, and its compressibility factor (Z) is still 0.9. Therefore, k1 (avg) still equals to 1.35, and Z1 (avg) still equals to 0.95. No further trial is required.

Hs_1 (required isentropic head, feet) $= 1545(0.95)(100+460)(2.86^{0.2593}-1)/(4*0.2593) = 248204.7$ [Eq. (2a)]
GHP1 $= 36.6*248204.7/(33000*0.75) = 366.9$ hp   [Eq. (4)]
BHP1 $= $ GHP1 $+$ GHP1$^{0.4} = 366.9+10.6 = 377.5$ hp   [Eq. (11)]

For second stage compressor:

Ps_2 (2nd stage suction pressure, psia) $= 185-5 = 180.0$; assume 2nd stage gas inlet temperature (Ts_2) $= 120°F$
Pd_2 (2nd stage discharge pressure, psia) $= 2.86*180 = 514.9$ [Eq. (2b)]

Assume at 2nd stage compressor inlet, gas isentropic exponent is 1.4, and its compressibility factor is 0.9.

Calculate discharge temperature using Eq. (4).

Td_2 (2nd stage discharge temperature, °R) $= (120+460)+(120+460)(2.86^{((1.4-1)/1.4)}-1)/0.75 = 850.8°R = 390.8°F$

Assume at this discharge conditions, gas isentropic exponent (k) is 1.3, and its compressibility factor (Z) is 0.8. The average isentropic exponent and compressibility factor for the 2nd stage compression are as follows:

$K2$ (avg) $= (1.4+1.3)/2 = 1.35$; $Z2$ (avg) $= (0.9+0.8)/2 = 0.85$;   Let $kx1 = (k1-1)/k1 = (1.35-1)/1.35 = 0.2593$

Recalculate Td_2 (2nd stage discharge temperature, °R):

$Td_2 = (120+460)+(120+460)(2.86^{((1.35-1)/1.35)}-1)/0.75 = 822.2°R = 362.2°F$

Assume at this discharge conditions, gas isentropic exponent (k) is still 1.3, and its compressibility factor (Z) is still 0.9. Therefore, k1 (avg) still equals to 1.35, and Z1 (avg) still equals to 0.95. No further trial is required.

Hs_2 (required isentropic head, feet) $= 1545(0.85)(120+460)(2.86^{0.2593}-1)/(4*0.2593) = 230009.2$ [Eq. (2a)]
GHP2 $= 36.6*230009.2/(33000*0.75) = 340.0$ hp   [Eq. (4)]
BHP2 $= $ GHP2 $+$ GHP2$^{0.4} = 340.0+10.3 = 350.3$ hp   [Eq. (11)]

Total BHP $= $ BHP1$+$BHP2 $= 727.8$ hp

Example 3 Rating (check process capacity) of an existing compressor for new service: From hydraulic study, we will know the required compressor head or differential pressure.
Depending on whether the gas in new service is the same type of gas designed for the existing compressor or not, we can have following two approaches for compressor rating.

(1) If the gas in new service is the same used to design the existing compressor, only flow rate is changed.

For centrifugal compressor, compressor performance curves such as Fig. 7a-c can be used to check whether the existing compressor can be used to handle the new gas flow rate or horsepower. If the new horsepower is not available on the curve, it can be calculated using Eq. (2a-c to 3) and Eq. (11). Compressor discharge temperature can be calculated using Eq. (4).

For reciprocating compressor, equations in section V.4 can be used to check compressor capacity at new service. The new horsepower can be calculated using Eq. (2a-c to 3) and Eq. (11). Compressor discharge temperature can be calculated using Eq. (4).

For rotary compressor, its new service is better to be checked by compressor manufacturer or compressor engineer.

(2) If the gas in new service is different from the one designed for the existing compressor, it is better to consult compressor manufacturer or compressor engineer to check the adequacy of an existing compressor for the new service (for centrifugal compressor, reciprocating compressor, or rotary compressor).

Chapter 4 Heat Exchanger

I. Comment

I.1 After study this chapter, reader should understand the basic features of double-pipe heat exchanger, shell and tube heat exchanger, and air cooler, and how to select a new heat exchanger among these three types of heat exchanger. Design or rating of an existing heat exchanger should rely on a computer software, such as HTRI computer programs, due to the complicated calculation of overall heat transfer coefficient and pressure drop.

II. Exercise

1. What is the purpose of using a heat exchanger?
2. Name at least 5 process information should be provided on a heat exchanger data sheet.
3. What is a heating or cooling curve?
4. Name three flow patterns usually found in heat exchangers.
5. Which equation should be used to calculate heat transfer duty of a heat exchanger?
6. Name one code used often for a shell and tube heat exchanger.
7. Which equation should be used to calculate overall heat transfer coefficient for a heat exchanger?
8. How to improve overall heat transfer coefficient in a heat exchanger?
9. How to reduce pressure drop in a heat exchanger?
10. What cause a heat exchanger to foul?
11. Name three major types of a heat exchanger. What is the difference among them?
12. Name major components of a double-pipe heat exchanger.
13. Name major components of a shell and tube heat exchanger.
14. Name major components of an air cooler.

III. Example

Example 1 A hydrocarbon liquid stream at 25,000 lb/hr is cooled from 150°F to 100°F. Its average heat capacity over this temperature range is 0.8 btu/lb-°F. Cooling water is used to cool down this stream. It is heated up from 80°F to 105°F. (1) What is the heat transfer duty for this service, in btu/hr? (2) What is the required cooling water rate in lb/hr? (3) Assume double-pipe heat exchanger is used for this service. Its overall heat transfer coefficient (U) is 100 btu/hr-ft2-°F (Note 1). What is the required heat transfer area, in ft2? (4) What type of heat exchanger should be used?

(1) Total heat transfer duty (Q), btu/hr = 25,000 * 0.8 * (150 – 100) = 1,000,000   [Eq. (2)]

(2) Required cooling water rate (Wc), lb/hr = Q / ((105 – 80) (1)) = 40, 000   [Eq. (3)] [Cpc = 1 btu/lb-°F]

(3) Required heat transfer area (A), ft2 = Q / (U*LMTD*F)   [Eq. (1)]

dT1 = 150 -105 =45°F, dT2 = 100 – 80 = 20°F, LMTD = (45 - 20)/ ln(45/20) = 30.83°F   [Eq. (5)]

For double-pipe heat exchanger, F = 1, since it is a countercurrent flow heat exchanger.

A, ft2 = 1,000, 000 / (100*30.83*1) = 324.4

(4) UA = 32,440 < 100,000 – Double-pipe heat exchanger can be used for this service. The assumption of using a double-pipe heat exchanger is correct.

Notes:

---

1. Overall heat transfer coefficient of a double-pipe heat exchanger is calculated based on selected tube size (diameter), number of tubes, and tube length. It is usually calculated by a computer software. Make sure unit's shell and tube side pressure drops are acceptable.

2. In this example, the cold stream outlet temperature is higher than the hot stream outlet temperature (105°F > 100°F). There is temperature crossover for this unit. If shell and tube heat exchanger is used likely two shells will be required.

3. 1 gpm cooling water flow rate is 500 lb/hr. 40,000 lb/hr cooling water flow rate is 80 gpm.

Example 2 A hot liquid stream at 400,000 lb/hr is cooled from 250°F to 200°F using a cold liquid stream, which is heated from 160°F to 200°F. The average heat capacity for the hot stream is 0.5 btu/lb-°F, and is 0.8 btu/lb-°F for the cold stream. (1) What is the heat transfer duty for this service, in btu/hr? (2) What is the required cold stream flow rate in lb/hr? Assume using shell and tube heat exchanger (one shell, two tube passes), the overall heat transfer coefficient (U) for this service is 150 btu/hr-ft2-°F (Note). (3) What is the required heat transfer area, in ft2? (4) What type of heat exchanger should be used?

(1) Total heat transfer duty (Q), btu/hr = 400,000 * 0.5 * (250 – 200) = 10,000,000   [Eq. (2)]

(2) Required cold stream flow rate (Wc), lb/hr = Q / ((200 – 160) (0.8)) = 312,500   [Eq. (3)]

(3) Required heat transfer area (A), ft2 = Q / (U*LMTD*F)   [Eq. (1)]

dT1 = 250 -200 =50°F, dT2 = 200 – 160 = 40°F, LMTD = (50 - 40)/ ln(50/40) = 44.81°F   [Eq. (5)]

For a shell and tube heat exchanger (one shell, two tube passes), F = 0.8. [The value of F can be found from a LMTD correction chart for a shell and tube heat exchanger with one shell and two tube passes. This chart is available from TEMA or other heat exchanger books.]

A, ft2 = 10,000, 000 / (150*44.81*0.8) = 1859.5

(4) UA = 278,929 > 100,000 – The assumption of using a shell and tube heat exchanger is correct.

Note: Overall heat transfer coefficient of a shell and tube heat exchanger is calculated based on selected tube size (diameter), number of tubes, tube length, number of tube passes, baffle space, and baffle cut. It is usually calculated by a computer software. Make sure unit's shell and tube side pressure drops are acceptable.

Example 3 An air cooler is used to cool a liquid stream at 200,000 lb/hr from 200°F to 150°F, and the air is heated from 100°F to 140°F. The average heat capacity for this hot stream is 0.5 btu/lb-°F. (1) What is the heat transfer duty for this air cooler, in btu/hr? (2) What is the required air flow rate in lb/hr? Assume the overall heat transfer coefficient (U) is 100 btu/hr-ft2-°F based on bare tube surface area (Note). (3) What is the required heat transfer area, in ft2?

(1) Total heat transfer duty (Q), btu/hr = 200,000 * 0.5 * (200 – 150) = 5,000,000   [Eq. (2)]

(2) Required air flow rate (Wc), lb/hr = Q / ((140 – 100) (0.25)) = 500,000   [Eq. (3)] [Cpc =0.25 btu/lb-°F]

(3) Required heat transfer area (A), ft2 = Q / (U*LMTD*F)   [Eq. (1)]

dT1 = 200 - 140 =60°F, dT2 = 150 – 100 = 50°F, LMTD = (60 - 50)/ ln(60/50) = 54.85°F   [Eq. (5)]

For air cooler, F = 0.985. [F is from a LMTD correction chart for air cooler, which can be found in GPSA Engineering Data Book (Section 10) or other air cooler books.]

A, ft2 = 50,000, 000 / (100*54.85*0.985) = 925.5

Note: Overall heat transfer coefficient of an air cooler is calculated based on selected tube size (diameter), number of tubes, tube length, and number of tube passes. It is usually calculated by a computer software. Make sure unit's tube side pressure drop is acceptable.

Example 4 Rate the performance of an existing double-pipe heat exchanger, a shell and tube heat exchanger, and an air cooler for a new service.

It is better to use a computer software to rate these existing heat exchangers, because of the complicated calculation of overall heat transfer coefficient and pressure drop. The target is to check (1) whether the existing heat exchanger can provide the required duty, and (2) whether the pressure drops at shell and tube side are acceptable.

For air cooler, we can increase the air face velocity through the unit to increase its duty to the limit of its fan driver's maximum power. Beyond this limit, a bigger fan driver is required. Maximum air face velocity is affected by number of tube rows used. In general, this velocity is 900 ft/min for 3 tube rows unit, 700 ft/min for 4 tube rows unit, 600 ft/min for 5 tube rows unit, and 500 ft/min for 6 tube rows unit. Consult manufacturer or an air cooler engineer for final decision. Other solution to increase duty of an air cooler is to replace its tube bundles.

Chapter 5 Vessel

I. Comment

I.1 After study this chapter, reader should understand the basic features of a vessel, how to size a liquid surge drum, a vapor/liquid separator, and to provide a preliminary liquid/liquid separator size. Reader should also be able to rate an existing surge drum or vapor/liquid separator for a new service.

I.2 For section IV.1 optimum liquid surge drum sizing (based on vessel weight): Usually among a set of vessel diameters (for a fixed surge volume), a vessel has smallest diameter and longest length will have the least weight. In reality, most vessel design engineer limits final vessel L/D ratio to 3 to 5.

I.3 Sizing vapor/(light) liquid/(heavy) liquid separator : This is a three phase separator. The light liquid is usually a hydrocarbon liquid, and the heavy liquid is usually water. Examples of three phase separator are (1) separate the well head associated gas, and (2) cold separator in a hydrotreater unit. For these service, usually mist eliminator is used to enhance the vapor and liquid separation using vertical vessel (see section IV.2.2a) or horizontal vessel (see section IV.2.2b).

For three phase separator, methods outlined in section IV.2.2a-b need to be extended to accommodate the heavy liquid phase. For a vertical vessel, a vessel section should be added for the heavy liquid phase, see Figure 10. For a horizontal vessel, a weir and a boot should be added for the light liquid phase to overflow the weir and the heavy liquid phase to accumulate in a boot, see Figure 11a. An alternate is to add a layer for the heavy liquid phase without using boot and weir, see Figure 11b. In general, this alternate scheme will cost less, since weir and boot are not required and vessel support pad can be shorter. For the scheme shown in Figure 10 and Figure 11b, consult instrument engineer about the distance from light liquid outlet nozzle to the high high interface liquid level (HHILL). Light liquid and heavy liquid surge volume can be estimated based on section IV.3 for preliminary sizing, or per project instruction or consult a vendor for final sizing.

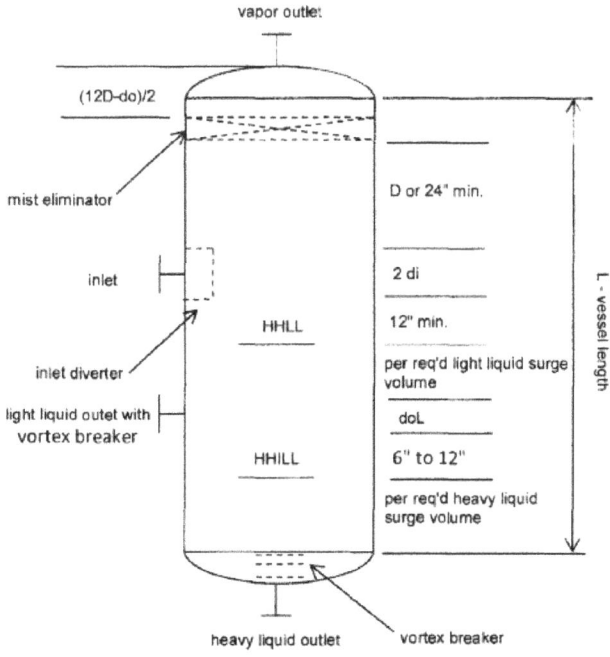

di - inlet nozzle diameter; do - vapor outlet nozzle diameter.
doL - light liquid outlet nozzle diameter.

Figure 10 A typical vertical vapor/liquid/liquid separator using mist eliminator.

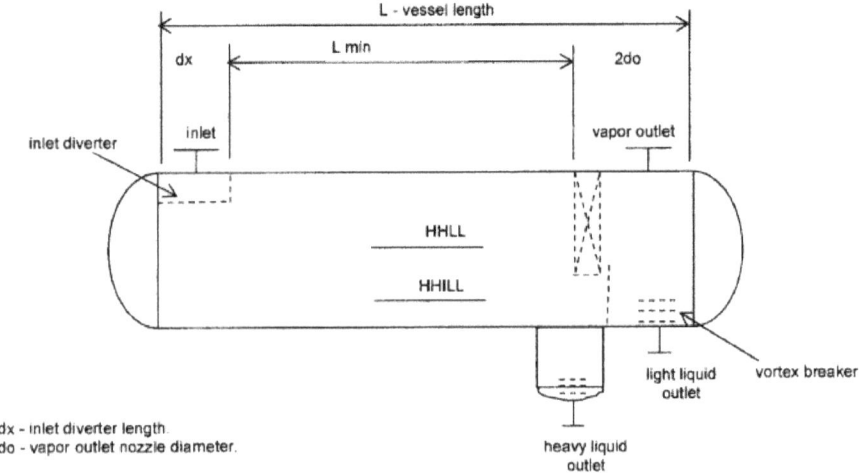

Figure 11a A typical horizontal vapor/liquid/liquid separator using mist eliminator and boot.

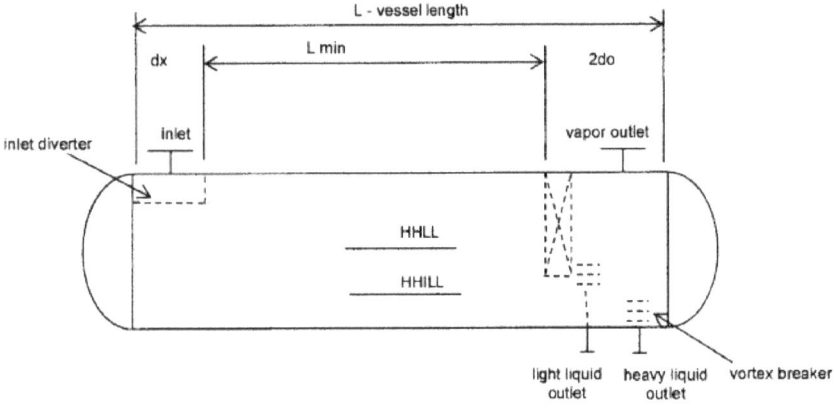

Figure 11b A typical horizontal vapor/liquid/liquid separator using mist eliminator.

Figure 11b assumes that inlet diverter length (dx) starts from vessel seam or tangent line. Consult vessel engineer the actual location of inlet diverter location and adjust vessel length, if necessary. This comment is also applied to Figure 2b and Figure 8b.

II. Exercise

1. What is the purpose of using a vessel?
2. Name at least 5 process information should be provided on a vessel data sheet.
3. Name at least three basic components of a vessel.
4. How to determine the design pressure and temperature of a vessel?
5. Which equation should be used to calculate vessel wall thickness?
6. Vessel costs depend on its weight. What factors will affect vessel's weight?
7. How to size a liquid surge drum?
8. How to size a vapor/liquid separator?
9. Which vapor/liquid separator sizing method will provide a smaller vessel?
10. How many types of mist eliminator are used in plant? Which two are commonly used and discussed in this book?
11. Which equation should be used to calculate allowable vapor velocity (va) using mist eliminator for vapor/liquid separator sizing?
12. How to size a liquid/liquid separator?
13. Which liquid/liquid separator sizing method will provide a smaller vessel?

14. What are the advantages of using a vertical vapor/liquid separator?
15. What are the advantages of using a horizontal vapor/liquid separator?

III. Example

Example 1 A liquid feed is fed to a surge drum at 100 gpm at 50 psig, 100°F. Required surge time is 5 minutes. (1) What will be the design pressure and temperature of this vessel. (2) Design this surge drum as a vertical or a horizontal vessel. This surge drum has HHLL and LLLL, and its corrosion allowance is 0.125 inch.

(1) Select vessel's design pressure and temperature: Per section III.6, they are selected as follows:

Design pressure (psig) = 50 +25 = 75; Design temperature (°F) = 100 + 50 = 150

Note: Assume 50 psig is the maximum operating pressure, and 100°F is the maximum operating temperature.

(2) Sizing:

Liquid surge volume (VL, ft3) = (100/7.4805)*(5+3*2) = 147.0 [Eq. (2a)]
Vessel cross sectional area (AT, ft2) = 147 /(0.8*L) = 183.8/L [Eq. (2b)]
Vessel diameter (D, ft) = 2*(AT/π)^0.5 [Eq. (2c)]

Following table is generated based on above equations. Let L start at 10 feet. Assume vessel joint efficiency (E) is 0.85 and total nozzle weight is 300 lbs.

L, ft	AT, ft2	D, ft	(selected)	L/D	dts, in.	(selected)	dth, in.	(selected)	Ws, lb	Wh, lb	WT, lb
10	18.38	11.70	12	0.833	0.443	0.5	0.443	0.5	7972	3177	14325
15	12.25	7.80	8	1.875	0.337	0.375	0.337	0.375	6054	1059	8172
20	9.19	5.85	6	3.333	0.284	0.3125	0.284	0.3125	5095	496	6088
25	7.35	4.68	5	5.000	0.258	0.25	0.257	0.25	4296	276	4847

Final vessel size is either 6 feet ID x 20 feet long or 5 feet ID x 25 feet long for vertical or horizontal design.

Example 2 A vapor/liquid stream is fed to a separator for separation at 50 psig, 100°F. Vapor flow rate is 1000 lb/hr. Its density is 0.5 lb/ft3, and its viscosity is 0.015 cp. Liquid flow rate is 100 gpm. Its density is 40 lb/ft3. Required liquid surge time is 3 minutes. Allow another 3 min for LLL to LLLL and no surge time for HLL to HHLL. Design this separator using (1) gravity settling method, (2) mist eliminator method.

(1) Using gravity settling method: to remove liquid particles 150 microns or larger in vapor.

1. First, calculate liquid particle settling velocity:

Settling velocity of 150 micron liquid particle (vs, ft/sec) = 0.01186 (150 (40-0.5)/(0.5C))^0.5 [Eq. (3a)]

Drag coefficient (C) is calculated as follows:

Ar = 2.517(10^-9) (150^3)(40-0.5)(0.5)/(0.015^2) = 745.66 [Eq. (3b)]
C = (432/ Ar) (1+0.047 (Ar^(2/3)))+0.517/(1+154 (Ar^(-1/3))) = 2.847 [Eq. (3c)]
Vs = 0.765 ft/sec

2. Calculate vapor design volumetric flow (QV, ft3/hr):

QV = 10000/0.5 = 20000 ft3/hr

3. For vertical vapor/liquid separator (section IV.2.1a):

Vessel cross sectional area (AT, ft2) = 20000 / (3600*0.765) = 7.261 [Eq. (4a)]
Vessel (inside) diameter (D, ft) = 3.04

Calculate liquid surge volume (VL, ft3).

VL = (100/7.4805) (3+3) = 80.21 ft3

Following table can be generated, for vessel diameter at 3, 3.5, and 4 feet, assume inlet nozzle dia. (di)=6 inch.

AT, ft2	D, ft	(selected)	ATf, ft2	VL, ft3	LL, ft	(selected)	T/T, ft	L/D
7.261	3.041	3.0	7.069	80.21	11.35	11.5	16.5	5.5
		3.5	9.621	80.21	8.34	8.5	14.0	4.0
		4.0	12.566	80.21	6.38	6.5	12.5	3.1

The final vessel size selection is 3.5 feet ID x 14 feet, since its diameter is little bit larger than required and it will allow liquid particles little bit less than 150 microns to settle down in the vessel (better separation).

4. For horizontal vapor/liquid separator (section IV.2.1b):

Vapor velocity (v, ft/sec) = 3 (0.765) = 2.295 [Eq. (4g), assume L/D=3]
Vapor cross sectional area (AV, ft2) = 20000 / (3600*2.295) = 2.42 [Eq. (4h)]

Let L=15 feet. Liquid cross sectional area (AL, ft2) = 80.21/15 = 5.35 ft2 [Eq. (4j)]

Vessel cross sectional area (AT, ft2) = 2.42+5.35 = 7.77 [Eq. (4k)]
Vessel diameter (D, ft) = 2 (7.77/π)^0.5 = 3.145 [Eq. (4m)]

For vessel length (L, ft) equals to 15, 17.5, and 12.5 feet, following table is generated.

AV, ft2	L, ft	VL, ft3	AL, ft2	AT, ft2	D, ft	(selected)	L/D	Lmin
2.42	15.0	80.21	5.35	7.77	3.145	3.0	5.0	11.0
2.42	17.5	80.21	4.58	7.00	2.986	3.0	5.8	11.0
2.42	12.5	80.21	6.42	8.84	3.354	3.5	3.6	12.5

Lmin for the last case is calculated by following equation:

Lmin = 3.5 * (2.295 / 0.765) +1+1 = 12.5 [Eq. (4i); let HV=D]

The final vessel size selection is 3.5 feet ID x 12.5 feet, for L/D < 5.

(2) Using mist eliminator method: Assume using wire mesh type mist eliminator.

1. For a vertical vapor/liquid separator (section IV.2.2a):

K = 0.35 ft/sec; allowable vapor velocity (va, ft/sec) = 0.35 [(40-0.5)/0.5]^0.5 = 3.11 [Eq. (5b)]

Vessel cross sectional area (AT, ft2) = 20000 / (3600*3.11) = 1.786 [Eq. (4a)]

Following table can be generated, for vessel diameter at 2, 3, and 3.5 feet, assume di=6 inch.

AT, ft2	D, ft	(selected)	ATf, ft2	VL, ft3	LL, ft	(selected)	T/T, ft	L/D
1.786	1.508	2.0	3.142	80.21	25.53	26	30.0	15
		3.0	7.069	80.21	11.35	11.5	16.5	5.5
		3.5	9.621	80.21	8.34	8.5	14.0	4.0

The final vessel size selection is 3.5 feet ID x 14 feet, for L/D < 5.

2. For horizontal vapor/liquid separator (section IV.2.2b):

K = 0.42 ft/sec; allowable vapor velocity (va, ft/sec) = 0.42 [(40-0.5)/0.5]^0.5 = 3.73   [Eq. (5b)]
Vapor cross sectional area (AV, ft2) = 20000 / (3600*3.73) = 1.49   [Eq. (4h)]

Let L=15 feet. Liquid cross sectional area (AL, ft2) = 80.21/15 = 5.35 ft2   [Eq. (4j)]

AT = 1.49+5.35 = 6.84 ft2   [Eq. (4k)]; Vessel diameter (D, ft) = 2 (6.84/π)^0.5 = 2.95   [Eq. (4m)]

For vessel length (L, ft) equals to 15, 17.5, and 12.5 feet, following table is generated.

AV, ft2	L, ft	VL, ft3	AL, ft2	AT, ft2	D, ft	(selected)	L/D
1.49	15.0	80.21	5.35	6.84	2.950	3.0	5.0
1.49	17.5	80.21	4.58	6.07	2.780	3.0	5.8
1.49	12.5	80.21	6.42	7.90	3.173	3.5	3.6

The final vessel size selection is 3.5 feet ID x 12.5 feet, for L/D < 5.

Notes:
1. For this example, the controlling sizing phase is liquid. Therefore, either method (gravity settling or mist eliminator) provides about same size of vessel. For vapor phase controlling case, mist eliminator method will provide a smaller vessel.
2. For an alternate vessel design using mist eliminator method, designer can use a smaller diameter vessel section for vapor, and a larger diameter vessel section for liquid.

Example 3 An amine/oil mixture at 100 psig, 80°F is to be separated in a vessel. The total flow rate is 100 gpm, with 75 gpm oil and 25 gpm amine. Design this separator using (1) surge volume method, (2) coalescer pad method.

(1) Using surge volume method (see section IV.3.1):

Per Table 1, required surge time is 30 min.

Total separator volume (VT, ft3) = 100*30/7.4805 = 401.0   [Eq. (6a)]

For vessel length (L, ft) = 15. Vessel cross sectional area (AT, ft2) = 401/15 = 26.74   [Eq. (7a)]

Vessel diameter (D, ft) = 2 (26.74/π)^0.5 = 5.83   [Eq. (7b)]

For vessel length 15, 20, and 25 feet, vessel diameter is calculated in following table.

L, ft	AT, ft2	D, ft	(selected)	L/D
15.0	26.74	5.83	6.0	2.50
20.0	20.05	5.05	5.0	4.00
25.0	16.04	4.52	4.5	5.56

The final vessel size selection is 5.0 feet ID x 20 feet for either vertical or horizontal vessel, for L/D < 5 and smaller diameter.

(2) Using coalesce pad method (see section IV.3.2):

1. For horizontal separator: keep flow rate at average recommended rate, 4.5 ft/min. (Note 1)

AT = 100/(7.4805*4.5) = 2.97 ft2; D = 1.94 ft, use 2 ft; for L/D=3, vessel length (L, ft) = 6

2. For vertical separator: keep flow rate at average recommended rate, 1 ft/min. (Note 1)

AT = 100/(7.4805*1.0) = 13.37 ft2; D = 4.13 ft, use 4.5 ft; for L/D=3, vessel length (L, ft) = 13.5 (Note 1)

Notes:
1. For conservative sizing, using the smaller recommended flow rate and L/D ratio at 3.
2. Liquid/liquid separator sizing methods presented are preliminary. For final sizing, consult a vendor or a specialist.

Example 4 How to rate the existing surge drum or vapor/liquid separator for a new service?

For an existing surge drum, calculate the surge time for the new service. Then, decide whether the surge time is acceptable. For an existing vapor/liquid separator, calculate the required vapor and liquid volume for the new service and to see whether they are smaller than the existing vessel or not. If it is, the existing vessel is adequate for the new service.

Chapter 6 Line Sizing

I. Comment

I.1 After study this chapter, reader should be able to size a single phase vapor or liquid line, two phase vapor/liquid line, and gravity flow line, and to provide preliminary slurry line size. Reader should also be able to check the adequacy of an existing line for new service in single phase vapor or liquid flow, two phase flow, or gravity flow.

I.2 Another source for two phase pressure drop calculation is Engineering Data Book, Section 17, Fluid Flow and Piping, p. 17-21, Gas Processors Suppliers Association, Tulsa, Oklahoma; 12th edition, 2004.

II. Exercise

1. What is the purpose of line sizing?
2. Name 4 types of line which are discussed in this chapter.
3. Name at least 5 factors of line sizing.
4. How to perform preliminary single phase line sizing?
5. Which equations should be used to calculate pressure drop in a single phase line?
6. Which equations should be used to calculate friction factor in a single phase line?
7. Name at least three flow patterns for a two phase flow line in horizontal and vertical direction.
8. How to perform preliminary two phase (vapor/liquid) line sizing?
9. Which two methods are discussed to calculate pressure drop in a two phase (vapor/liquid) line?
10. What is the driving force for a gravity flow line?
11. There is a range of line sizes can be used for a gravity flow line. True or False.
12. Which equation should be used to calculate the largest line size for a gravity flow line?
13. Name four flow regions for a horizontal slurry line
14. How to calculate pressure drop for a pipe fittings?
15. How to do the final line sizing?

III. Example

Example 1 A liquid line is operated at following inlet conditions: 500 gpm, 50 psig, 100°F, density at 40 lb/ft3, and viscosity at 0.5 cp. Provide preliminary line sizing of this line based on (1) maximum line velocity at 3 ft/sec, (2) allowable line pressure drop at 2 psi/100 ft pipe. Assume line roughness is 0.00015 feet.

(1) Preliminary sizing of this line based on maximum line velocity at 3 ft/sec:

Q=500 gpm=500/7.4805/60=1.11 ft3/sec

Choose several line sizes and calculate its line velocity in following table:

line size, inch	line inside diameter, inch	line cross sectional area, ft2	line velocity, ft/sec
6"s40	6.065	0.2006	5.6
8"s40	7.981	0.3474	3.2
10"s40	10.02	0.5475	2.0

Since the maximum line velocity is 3 ft/sec, the proper line size should be 10" sch. 40.

(2) Preliminary sizing of this line based on allowable line pressure drop at 2 psi/100 ft pipe:

Choose several line sizes and calculate its line pressure drop in following table:

line size, inch	line inside diameter, inch	line velocity, ft/sec	Reynolds number	friction factor	line pressure drop, psi/100ft
4"s40	4.026	12.6	503299	0.0174	3.561
6"s40	6.065	5.6	334095	0.0169	0.444

Since the line allowable pressure drop is 2 psi/100ft, the proper line size should be 6" sch. 40, with line pressure drop less than 2 psi/100ft.

Example 2 A vapor line is operated at following inlet conditions: 10,000 lb/hr, 50 psig, 100°F, density at 0.5 lb/ft3, and viscosity at 0.01 cp. Line length is 100 feet. Provide preliminary line sizing of this line based on (1) maximum line velocity at 50 ft/sec, (2) allowable line pressure drop at 0.5 psi/100 ft pipe. Assume line roughness is 0.00015 feet.

(1) Preliminary sizing of this line based on maximum line velocity at 50 ft/sec:

w= 10000 lb/hr = 10000/3600/0.5 = 5.56 ft3/sec

Choose several line sizes and calculate its line velocity in following table:

line size, inch	line inside diameter, inch	line cross sectional area, ft2	line velocity, ft/sec
4"s40	4.026	0.0884	62.8
6"s40	6.065	0.2006	27.7

Since the maximum line velocity is 50 ft/sec, the proper line size should be 6" sch. 40.

(2) Preliminary sizing of this line based on allowable line pressure drop at 0.5 psi/100 ft pipe:

Choose several line sizes and calculate its line pressure drop in following table:

line size, inch	line inside diameter, inch	line velocity, ft/sec	Reynolds number	friction factor	line pressure drop, psi/100ft
4"s40	4.026	62.8	1568721	0.0167	1.062
6"s40	6.065	27.7	1041331	0.0157	0.1285

Since the allowable line pressure drop is 0.5 psi/100ft, the proper line size should be 6" sch. 40, with pressure drop less than 0.5 psi/100ft.

Check acceleration factor (AC): Line length=100 ft. For 6" sch. 40 line, its cross sectional area (A, ft2)=0.2006

P1= 64.7 psia; P2=64.7-0.128*100/100=64.572 psia; Pavg=(64.7+64.572)/2=64.636 psia [Eq. (4d)]
Mass flux (G, lb/sec-ft2) = 5.56/0.2006=13.847
AC = 13.847*27.7/(144*32.174*64.636)=0.00128

Final line pressure drop (DP, psi/100ft) = 0.1285/(1-0.00128) =0.1287<0.5

Selected line size 6" sch. 40 is correct.

Example 3 A two phase (vapor/liquid) flow line is operated at following inlet conditions: at 50 psig, 100°F, vapor flow rate is 10,000 lb/hr, vapor density at 0.5 lb/ft3, and vapor viscosity at 0.01 cp; liquid flow rate is 500 gpm, liquid density at 40 lb/ft3, liquid viscosity at 0.5 cp, and its surface tension is 15 dyne/cm. Line length is 100 feet. Provide preliminary line sizing of this line based on (1) maximum line velocity at 50 ft/sec, (2) allowable line pressure drop at 0.5 psi/100 ft pipe. Assume line roughness is 0.00015 feet.

Vapor weight flow (wv, lb/sec)=10000/3600=2.78
Liquid weight flow (wL, lb/sec)= 500*40 / (60*7.4805) = 44.56
Total flow: wm =2.78+44.56=47.34 lb/sec [wt. flow]= 47.34/7.1 = 6.67 ft3/sec [volume flow]

Mixture (vapor/liquid) density (ρ mix, lb/ft3) = (2.78+44.56)/(2.78/0.5+44.56/40) = 7.1   [Eq. (8)]

(1) Preliminary sizing of this line based on maximum line velocity at 50 ft/sec:

Choose several line sizes and calculate its line velocity in following table:

line size, inch	line inside diameter, inch	line cross sectional area, ft2	line velocity, ft/sec
4"s40	4.026	0.0884	75.4
6"s40	6.065	0.2006	33.2

Since the maximum line velocity is 50 ft/sec, the proper line size should be 6" sch. 40.

(2) Preliminary sizing of this line based on allowable line pressure drop at 0.5 psi/100 ft pipe:

Two phase flow line pressure drop can be calculated by following two methods, see section III.2.2.2:

1. Homogeneous flow method (Dukler no slip method)

Liquid volume flow rate (QL, ft3/sec)=44.56/40=1.11
vapor volume flow rate (Qv, ft3/sec)=2.78/0.5=5.56
Liquid volume fraction (x)=1.11/(1.11+5.56)=0.167   [Eq. (15a)]

No slip density (ρm_{ns}, lb/ft3)=40*0.167+0.5*(1-0.167)=7.1   [Eq. (15b)]
No slip viscosity (μm_{ns}, cp)=0.5*0.167+0.01*(1-0.167)=0.092   [Eq. 915c)]

Choose several line sizes and calculate its line pressure drop as following table:

line size, inch	line inside diameter, inch	line velocity, ft/sec	Reynolds number	friction factor	line pressure drop, psi/100ft
8"s40	7.981	19.2	1468333	0.0148	0.627
10"s40	10.02	12.2	1169538	0.0144	0.196

Since allowable line pressure drop is 0.5 psi/100ft, the proper line size should be 10" sch. 40.

For 10" sch. 40 pipe: line inside diameter (D, ft)=10.02/12=0.835; cross sectional area (A, ft2)=0.5476

Check acceleration factor (AC): for a line of 100 ft.

Mass flux of the two phase flow (G, lb/sec-ft2) = 47.34/0.5476=86.45 [Eq. (16d)]
P1= 64.7 psia; P2=64.7-0.196*100/100=64.504 psia; Pavg=(64.7+64.504)/2=64.602 psia [Eq. (16s)]
AC = 86.45*12.2/(144*32.174*64.602)=0.003518 [Eq. (16r)]

Final line pressure drip (psi/100ft)= 0.196/(1-0.003518) =0.197<0.5 [Eq. (16q)]

Selected line size 10" sch. 40 is correct.

2. Dukler constant slip method:

Based on above calculation (Homogeneous flow method), selected line size is 10" sch. 40. We will use this line size to calculate line pressure drop using Dukler constant slip method.

Per above calculation: Liquid volume fraction (x)=0.167; No slip density (ρm_{ns}, lb/ft3)=7.1; No slip viscosity (μm_{ns}, cp)=0.092

No slip two phase flow velocity (vm, ft/sec) = (QL+Qv)/A=(1.11+5.56)/0.5476=12.18 [or use Eq. (16g)]

Find RL (constant slip liquid volume fraction in place inside the line) by iterated calculation as follows:

First trial is explained below: Let RL= x=0.167
Calculate Froude number (Fr) = 12.18^2/(32.174*0.835)=5.522 [Eq. (16f)]
Calculate constant slip viscosity (μm_{cs}, cp)=0.5*0.167+0.01*(1-0.167)=0.092 [Eq. (16e)]
Calculate mass flux (G, lb/sec-ft2)=(2.78+44.56)/0.5476=86.45 [Eq. (16d)]
Calculate Reynolds number (Re)=1488.16*0.835*86.45/0.092=1169588.6 [Eq. (16c)]
Calculate Z (a correlating factor)=(1169588.6)^(1/6)*(5.522^(1/8)/(0.167^0.25)=19.88 [Eq. (16b)]

Once Z is known, Hughmark's flow parameter (K) can be calculated as follows:

K=0.75545+0.003585*19.88-0.00001436*(19.88^2)=0.82104

Once K is known, RL is recalculated using Eq. (16k) as follows:

RL = 1-0.82104/(1+0.5/40*((2.78+44.56)/2.78-1))=0.3161

Once RL is known, the second trial can be started. Repeat this trial calculation until calculated RL is no longer changed. Following table shows how RL is calculated. [Note: G, ρm_{ns}, vm, and Froude number are constants in this calculation.]

Trial	RL	μm_{cs}, cp	Re	Z	K
1	0.1670	0.092	1169588.6	19.88	0.821042
2	0.3161	0.165	651476.1	18.03	0.815426
3	0.3208	0.167	642542.5	17.99	0.815299
4	0.3209	0.167	642343.1	17.99	0.815296
5	0.3209	0.167	642338.6	17.99	0.815296

The final RL =0.3209.

Calculate constant slip density (ρm_{cs}, lb/ft3) using Eq. (16a):

ρm_{cs} =40*(0.167^2/0.3209)+0.5*(1-0.167^2)/(1-0.3209)=4.193

Constant slip Reynolds number is calculated as follows:

Re_{ns} = 1488.16*0.835*86.45/0.092=1169588.6 [Eq. (16o)]
Re_{cs} =1160588.6*4.193/7.1=691019.7 [Eq. (16n)]

Use Re_{cs} to calculate uncorrected friction factor (fx) per Eq. (7a). Calculated fx=0.0149. Calculate friction factor correction factor (α) as follows:

Ln(x) = Ln(0.167)= 1.7896
α = 2.2826 [Eq. (16n)]

Two phase mixture friction factor (fm) =0.0149*2.2826=0.034
Line pressure drop (DP per 100 ft line)=0.034*4.193*100*12.18^2/(24*32.174*10.02)=0.2736 [Eq. (2)]

Check acceleration factor (AC): for a line of 100 ft.

P1= 64.7 psia; P2=64.7-0.2736*100/100=64.4264 psia; Pavg=(64.7+64.4264)/2=64.5632 psia [Eq. (16s)]

AC = 0.001858 [Eq. (16t)]

Final line pressure drip (psi/100ft)=0.2736/(100.001858)=0.2742<0.5 [Eq. (16q)]

Selected line size 10" sch. 40 is correct. This method is very complicated, and it is better to use a computer (such as MS Excel spreadsheet) to do the calculation.

Note: Eq. (16g) is per Hughmark's original article. If ρm_{ns} in Eq. (16g) is replaced by ρm_{cs}, it is found that RL=0.3145, and final line pressure drip is 0.2658 psi/100ft. This shows line pressure drop calculation is not sensitive to which density (no slip or constant slip) is used in Eq. (16g).

Example 4 Size a gravity flow line for liquid flow at 200 gpm. Liquid density is 40 lb/ft3 and viscosity is 0.5 cp. Available static head is 5 feet or 1.389 psi. Total line length is 100 feet and line roughness is 0.00015 feet.

There are a set of lines can be used for this service. The largest one is based on self-vent design. It is calculated as follows:

Self-vent drain line diameter (d, inch) = 0.9168*(200)^0.4 =7.63 [Eq. (17)]

Use 8"sch. 40 line, d=7.981 in.

Calculate line pressure drop for 3"s40, 4"s40, 6"s40, and 8"s40 line as following table:

Line size, inch	Line inside diameter, inch	Line velocity, ft/sec	Reynolds Number	Friction factor	Line pressure drop, psi/100ft
3" sch. 40	3.068	8.7	264183	0.0190	2.413
4" sch. 40	4.026	5.0	201320	0.0186	0.609
6" sch. 40	6.065	2.2	133638	0.0187	0.079
8" sch. 40	7.981	1.3	101555	0.0191	0.020

Total available pressure drop is 1.389 psi. Therefore, lines from 4" to 8" will work. Final selection is 6"s40.

Example 5 How to rate an existing line for new service?

For preliminary rating, calculate its pressure drop or velocity to see whether it is acceptable. For final rating, check whether the line loss is acceptable (to check whether there is enough pressure available for the line loss or not.).

Chapter 7 Control valve

I. Comment

I.1 After study this chapter, reader should know the basic features of globe and rotary control valve, and be able to size a single phase vapor or liquid control valve. Reader should also be able to check the adequacy of an existing control valve for a new single phase vapor or liquid service.

II. Exercise

1. What is the purpose of a control valve?
2. Name types of control valve used in industry by application, by fluid, and by hardware.
3. Name major components of a globe control valve.
4. Name major components of a rotary control valve.
5. What is inherent control valve performance?
6. What is installed control valve performance?
7. Which equation should be used to size a liquid control valve?
8. Show pressure profile inside a liquid control valve along its flow path?
9. For liquid control valve, what are flash and cavitation services? Which one will cause more damage to valve?
10. Which equation should be used to calculate choke flow pressure drop for a liquid control valve?
11. Which equation should be used to calculate maximum flow through a liquid control valve?
12. How to calculate a liquid control valve's Reynolds number?
13. Which equation should be used to size a vapor control valve?
14. Flow through vapor control valve is limit by sonic flow. How do we know sonic flow is reached?
15. Which equations should be used to calculate control valve pressure drop at other flow conditions?
16. What is the maximum and minimum opening for a globe and rotary control valve?
17. Name at least 5 process information should be provided on a liquid or vapor control valve data sheet.
18. How to select a globe or rotary control valve for a new service?

III. Example

Example 1 A 110 gpm (maximum flow) liquid stream flows through a control valve at 100 psig, 100°F. Liquid density is 40 lb/ft3. Its vapor pressure is 10 psia and critical pressure is 500 psia. The allowable pressure drop for control valve is 15 psi. The system pressure drop (excluding control valve) and static pressure drop is 10 psi. Normal flow is 100 gpm and minimum flow is 50 gpm. What type control valve and size should be used?

(1) Control valve sizing and selection:

1. Assume there is no reducer at control valve inlet and outlet. Piping geometry factor (Fp)=1.0

sp. gr.= 40/62.37=0.6413

Let subscript 1 for Normal flow and 2 for minimum flow. Q1 = 100 gpm; Q2 = 50 gpm

Control valve pressure drop at normal flow (dp1, psi) = 15+10 (1- (100/110)^2)=16.7   [Eq. (18a)]
Control valve pressure drop at minimum flow (dp2, psi) = 15+10 (1- (50/110)^2)=22.9   [Eq. (18a)]

Cv for maximum flow = 110/((15/0.6413)^0.5) = 22.75   [Eq. (1)]

For this Cv, a globe control valve should be adequate for the service. From Fisher Catalog 12, 1.5" Ed globe valve can be selected for this service, with max Cv at 35.8. Control valve % open at maximum flow is 77.5%, which is less than 80%. Assume control valve inlet and outlet piping are 1.5" sch. 80 piping.

For normal flow, Cv = 19.58, valve is 73.1% open. For minimum flow, Cv = 8.36, valve is 52.4%open, which is greater than 15%. Therefore, selected control valve is adequate for intended service.

2. Assume there is reducer, 2"x1.5", at control valve inlet and outlet. Fp is calculated by following equations:

Assume control valve inlet and outlet piping size is 2" sch. 80, its inside diameter (D1or D2, inch) is 1.939.
$kb1 = kb2 = 1- (1.5/1.939)^4 = 0.642$ [Eq. (4d-e)]
$k1 = 0.5 (1-(1.5/1.939)^2)^2 = 0.081$ [Eq. (4b)]; $k2 = 1.0 (1-(1.5/1.939)^2)^2 = 0.161$ [Eq. (4c)]
$ks = 0.161+0.081+0.358-0.358 = 0.242$ [Eq. (4a)]
$Fp = (1+0.242 (22.75/*1.5^2)^2/890)^{(-0.5)} = 0.986$ for max. flow [Eq. (3)]
$Cv = 22.75 / 0.986 = 23.06$. Valve is 78% open.

For normal flow, Fp = 0.990. Cv = 19.78. Valve is 73.4% open.
For minimum flow, Fp = 0.998. Cv = 8.38. Valve is 52.5% open.

(2) Check the valve Reynolds number as follows: Liquid control valve pressure recovery factor (FL) =0.84. Calculate Re based on minimum flow control valve sizing coefficient.

$Re = 17300*50*(0.84^2*8.38^2/(890*1.939^4)+1)^{0.25}/((0.5/0.6413)*(0.84*8.38)^{0.5}) = 418672$ [Eq. (9a)]

Since Re is greater than 40,000, flow in control valve is turbulent, and the sizing equations used are correct.

(3) Check choked flow:

Liquid control valve critical pressure ratio factor (FF) = $0.96-0.28*(10/500)^{0.5} = 0.92$ [Eq. (6)]
Minimum choked flow control valve pressure drop (dp choke, psi) = $(0.84^2) (114.7-0.92*10) = 74.4$ [Eq. (5)]

Since control valve pressure drop (15 psi) is less than dp choke, flow in this control valve is not choked.

Example 2 A 110 gpm liquid stream flows through a control valve at 100 psig, 100°F. This is the only flow for this case. Liquid density is 40 lb/ft3. Its vapor pressure is 100 psia and critical pressure is 500 psia. The allowable pressure drop for control valve is 25 psi. What type control valve and size should be used? There is no reducer at control valve inlet and outlet.

(1) Control valve sizing and selection:

$Cv = 110/((25/0.6413)^{0.5}) = 17.62$ [Eq. (1)]

For this Cv, globe control valve should be adequate for the service. From Fisher Catalog 12, 1.5" Ed globe valve can be selected for this service, with max Cv at 35.8. Control valve is 70.3% open, which is less than 80%.

(2) Check choked flow:

Liquid control valve pressure ratio (FF) = $0.96-0.28 (100/500)^{0.5} = 0.835$ [Eq. (6)]
Minimum choked flow control valve pressure drop (dp choke, psi)= $0.84^2 (114.7-0.835*100)= 22.0$ [Eq. (5)]

Since control valve pressure drop (25 psi) is greater than dp choke, flow in this control valve is choked.

(3) Check maximum flow through control valve (Q max):

$Q max = 17.62 (22.0/0.6413)^{0.5} = 103.3$ gpm [Eq. (7)]

For flow at 110 gpm, Cv should be as follows:

$Cv = 17.62 (110/103.3) = 18.77$; control valve is 71.9% open.

Example 3 A 110,000 lb/hr (maximum flow) vapor stream flows through a control valve at 100 psig, 100°F. Its molecular weight is 20. At valve inlet, its compressibility is 1.0 and its molar heat capacity (Cp, btu/mole-°F) is 10. The allowable pressure drop thru control valve is 5 psi. The system pressure drop excluding control valve is 10 psi. Normal flow is 100,000 lb/hr and minimum flow is 50,000 lb/hr. What type control valve and size should be used? There is no reducer at control valve inlet and outlet.

At maximum flow:

Fk = (10/(10-1.986)) / 1.4 = 0.8913 [Eq. (16)]; X = 5/114.7 = 0.0436 [Eq. (15a)] – for subsonic flow
XT = 0.36 [section V.2 for butterfly rotary valve]; Y = 1-0.0436/(3*0.8913*0.36) = 0.9547
Cv = 110000/(19.3*114.7*0.9547*(0.0436*20/(560*1))^0.5)=1319.1 [Eq. (13)]

For this large Cv, a rotary control valve is required. From Fisher Catalog 12, a 10" CV500 disc rotary valve can be used for this service. Use XT from the catalog to update the Y and Cv. Final values of XT, Y, and Cv are 0.2793, 0.9416, and 1337.4. For Cv at 1337.4, the valve disc is 70.1° open, between 60° and 90°.

Let subscript 1 for normal flow and 2 for minimum flow. W1=100000 lb/hr, W2=50000 lb/hr

Control valve pressure drop at normal flow (dp1, psi) = 5+10 (1- (100000/110000)^2)=6.7 [Eq. (18a)]
Control valve pressure drop at minimum flow (dp1, psi) = 5+10 (1- (50000/110000)^2)=12.9 [Eq. (18a)]

At normal flow:

Fk1 = (10/(10-1.986)) / 1.4 = 0.8913 [Eq. (16)]; X1 = 6.7/114.7 = 0.0587 [Eq. (15a)] – for subsonic flow
XT1 = 0.2979 [from Fisher Catalog 12, for 10" CV500]; Y1 = 1-0.0587/(3*0.8913*0.2979) = 0.9263
Cv1 = 100000/(19.3*114.7*0.9263*(0.0587*20/(560*1))^0.5)=1064.9 [Eq. (13)]   62.3° open.

At minimum flow:

Fk2 = (10/(10-1.986)) / 1.4 = 0.8913 [Eq. (16)]; X2 = 12.9/114.7 = 0.1128 [Eq. (15a)] – for subsonic flow
XT2 = 0.4811 [from Fisher Catalog 12, for 10" CV500]; Y2 = 1-0.1128/(3*0.8913*0.4811) = 0.9123
Cv2 = 50000/(19.3*114.7*0.9123*(0.1128*20/(560*1))^0.5)=390.1 [Eq. (13)]   35.8° open > 20°.

Selected control valve at maximum flow is between 60° and 90°, and at minimum flow is above 20°. Therefore, selected control valve is all right for the new service.

Example 4 A 50,000 lb/hr vapor stream flows through a control valve at 30 psig, 100°F. This is the only flow rate through the control valve. Vapor molecular weight is 20. At valve inlet, its compressibility is 1.0 and its molar heat capacity (Cp, btu/mole-°F) is 10. The allowable pressure drop thru control valve is 15 psi. What type control valve and size should be used? There is no reducer at control valve inlet and outlet.

Control sizing and selection:

Fk = (10/(10-1.986)) / 1.4 = 0.8913 [Eq. (16)]; X = 15/44.7 = 0.3356 [Eq. (15a)] – for subsonic flow
XT = 0.36 [section V.2 for butterfly rotary valve]; Y = 1-0.3356/(3*0.8913*0.36) = 0.6514 <0.667

Since Y is less than 0.667, we know flow thru control valve is sonic. Let Y=0.667, and
X=0.8913*0.36=0.3209 [Eq. (15b)] for sonic flow.
Cv = 50000/(19.3*44.7*0.667*(0.3209*20/(560*1))^0.5)=812.1 [Eq. (13)]

For this large Cv, a rotary control valve is required. From Fisher Catalog 12, a 8" CV500 disc rotary valve can be used for this service. Use XT from the catalog to update the Y and Cv. Final values of XT, Y, and Cv are 0.3557, 0.667, and 794.1. For Cv at 794.1, the valve disc is 68.4° open, between 60° and 90°.

Example 5 How to rate an existing control valve?

Calculate existing control valve's Cv under new operating conditions. Then, check whether its valve opening is acceptable. If it is, the existing valve is adequate for the new service. Otherwise, it is not, and a new valve is required.

Chapter 8 Pressure Relief Device

I. Comment

I.1 After study this chapter, reader should know the basic features of following pressure relief devices (PRD): spring-loaded pressure relief valve, pilot-operated pressure relief valve, rupture disc, and pin-actuated pressure relief device. Reader should be able to select and size a pressure relief device for a new service, either for vapor/liquid single phase relief or two-phase relief. Reader should also be able to check the adequacy of an existing pressure relief device for a new service.

I.2 PRD sizing equations used in this book are based API Standard 520, Sizing, Selection, and Installation of Pressure-relieving Devices in Refineries, Part I-Sizing and Selection, 8th edition, Dec. 2008.

II. Exercise

1. What is the purpose of a pressure relief device?
2. Name four types of pressure relief devices discussed in this book.
3. Which pressure relief devices are for reclosing service and which ones are for non-reclosing service?
4. What are the main differences in these four types of pressure relief devices?
5. What are the two commonly used spring-loaded pressure relief valves?
6. How to select between the two commonly used spring-loaded pressure relief valves?
7. What are the advantages of using pilot-operated PRV over spring-loaded PRV?
8. Name three major types of rupture disk.
9. What are the advantages of using pin-actuated PRD over rupture disk?
10. Describe the characteristics of a spring-loaded pressure relief valve.
11. Describe the characteristics of a pilot-operated pressure relief valve.
12. How to set a PRD's set pressure and overpressure?
13. What is the maximum equipment operating pressure for using the four types PRD?
14. What is contingency analysis?
15. Which equations should be used to calculate required orifice of a vapor pressure relief valve?
16. Which equation should be used to calculate required orifice of a liquid pressure relief valve?
17. Which equations should be used to calculate required orifice of a two-phase (vapor/liquid) pressure relief? Can these equations be used to calculate other relief cases?
18. What are the PRV inlet piping sizing criteria?
19. What are the PRV outlet piping sizing criteria?
20. How to select a PRD among the four types discussed in this book?

III. Example

Example 1 A vessel operating at 100 psig, 100°F. Its design pressure is 110 psig at 150°F. (1) What is the set pressure for its pressure relief valve? (2) If there are two pressure relief valves required to protect the vessel, what is the set pressure for the second pressure relief valve?

Since MAWP is not available, let vessel design pressure be the PRV set pressure. For nonfire or fire case:

(1) PRD set pressure, psig = 110*1.0 = 110   [per section IV, Table 1]
(2) 2nd PRD set pressure, psig = 110*1.05 = 115.5 psig   [per section IV, Table 1]

Example 2 During contingency analysis, it is found that a vapor stream will be relieved at 10000 lb/hr at 10% overpressure and 100°F. The PRD is a spring-loaded PRV. PRV set pressure is 110 psig. Relief vapor has following properties: mw (M) 20, compressibility factor (Z) 1.0, heat capacity ratio (k) 1.2. Assume backpressure correction factor (kb) = 1.0. What is the PRV size for (1) sonic relief, and (2) subsonic relief?

Per section VII.1, effective discharge coefficient (kd) = 0.975; kc = 1.0, since there is no rupture disc at PRV inlet; T = 100+460 =560°R; P1 = 110*1.1+14.7 = 135.7 (PRV inlet pressure, assume no inlet line.)
C = 520 (1.2 (2/(1.2+1))^((1.2+1)/(1.2-1))^0.5 = 337.24 [Eq. (7b)]

(1) For sonic relief:

PRV orifice (A, inch^2) = 10000 ((560*1) / 20)^0.5 / (337.24*0.975*1*1*135.7) = 1.19 [Eq. (7a)]
J orifice is required for this service.

(2) For subsonic relief: This will happen if PRV outlet pressure (P2) is greater than the critical pressure (Pc).

Pc = 135.7 (2/(1.2+1))^(1.2/(1.2-1)) = 76.6 psia [Eq. (6a)]; Assume P2 = 80 psia
r = 80/135.7 =0.59 [Eq. (8b)]
F2 = (1.2/(1.2-1)*0.59^(2/1.2)*(1-0.59^((1.2-1)/1.2))/(1-0.59))^0.5 = 0.715 [Eq. (8c)]
PRV orifice (A, inch^2) = 10000 ((560*1)/(20*135.7*(135.7-80)))^0.5/(735*0.327*0.975*1)=1.188 [Eq. (8a)]
J orifice is required for this service.

Example 3 During contingency analysis, it is found that a liquid stream will be relieved at 100 gpm at 10% overpressure and 100°F. The PRD is a pilot-operated PRV and it is certified. PRV set pressure is 110 psig. Relief liquid has following properties: liquid density 40 lb/ft3, viscosity (μ) 0.5 cp. PRV outlet pressure is atmospheric pressure.

Liquid relief load (QL, gpm) = 100; PRV inlet pressure (p1, psig) = 110*1.1 = 121 psig;
PRV outlet pressure (p2, psig) = 0; sp. gr. = 40/62.37 = 0.6413

Per section VII.2, effective discharge coefficient (kd) = 0.74; backpressure correction factor (kw) = 1.0, for pilot-operated PRV; kc = 1.0, since there is no rupture disc at PRV inlet; let viscosity correction factor (kv)=1.0

PRV orifice (A, inch^2) = 100 (0.6413/(121-0))^0.5/(38*0.74*1*1*1) = 0.259 [Eq. (9a)]

Update kv using Eq. (9b-c) and recalculate orifice area (A) by Eq. (9a). Final Re, kv, and A are 706926, 1.003, and 0.258 in^2. For orifice area at 0.258 in^2, F orifice is required for this service.

Example 4 During contingency analysis, it is found that a two-phase (vapor/liquid) stream will be relieved at 10% overpressure and 100°F. PRV set pressure is 110 psig. The total relief rate is 42083 lb/hr. The density of relief vapor is 0.452 lb/ft3, and for relief liquid, it is 40 lb/ft3 (at 10% overpressure). Vapor weight fraction is 0.2376. At PRV discharge at 20 psig, the density of relief vapor is 0.147 lb/ft3, and for relief liquid, it is 39 lb/ft3. Vapor weight fraction is 0.2614. Calculate required PRV orifice area.

First step, calculate constants based on given vapor/liquid density and vapor weight fraction as follows:

PRV inlet pressure (P1, psia)=110*1.1+14.7=135.7, ρv1, lb/ft3=0.452, ρL1, lb/ft3=40, x1=0.2376
For PRV discharge at 20 psig (P2, 34.7 psia): ρv2, lb/ft3=0.147, ρL2, lb/ft3=39, x2=0.2614

b = (0.2376-0.2614/(135.7-34.7)=-0.00024, a = 0.2376-(-0.00024)(135.7)=0.2696,
n =ln(135.7/34.7)/ln(0.452/0.147)=1.2141, c =(0.452)/(135.7)^(1.2141)=0.00792,
e =(40-39)/(135.7-34.7)=0.0099, d =40-0.0099*135.7=38.656

Second step, calculate following parameters at different pressure inside PRV: pressure (Pi, psia), vapor weight fraction (xi), vapor specific volume (vgi, ft3/lb), liquid specific volume (vfi, ft3/lb), $\sum$ [(xi * vgi + (1-xi) vfi]$_{avg}$ Δp (in (ft3/ lb)-psi), and G^2/9266 (=($\sum$$\Delta$p*vi,avg) / (xi * vgi + (1-xi) vfi)^2) (in (lb/sec-ft2)^2), for i=1 to m. List this parameters in a table as follows. The first two steps of calculation are explained below.

Let Δp = 2 psi. (This item is user's choice.)

For step 1 calculation (for the 1st row in the table):
P1=135.7 psia, x1=0.2376, vg1=2.2124, vf1=0.0250, v1=0.2376*2.2124+(1-0.2376)*0.0250=0.5447,
For this step, skip the calculation of the last two parameters.

For step 2 calculation (for the 2nd row in the table):
P2=135.7-2=133.7 psia, x2=0.2696-0.00024*133.7=0.2381,
vg2=1/(0.000792*133.7^(1/1.2141))=2.2396 ft3/lb, vf2=1/(38.6564+0.0099*133.7)=0.02501 ft3/lb,
v2 = 0.2381*2.2396+(1-0.2381)*0.02501=0.5522,
$\sum \Delta p*vi,avg$=0+2*(0.5447+0.5522)/2=1.09697, [Note: 0 is from 1st row.]
G^2/9266=1.09697/0.5522^2=3.5969

Repeat step 2 calculation for the rest steps. Following table can be constructed.

Calc. step(i)	Pi	xi	vgi	vfi	vi	$\sum \Delta p*vi,avg$	G2/9266
1	135.7	0.2376	2.2124	0.02500	0.5447		
2	133.7	0.2381	2.2396	0.02501	0.5522	1.09697	3.5969
3	131.7	0.2385	2.2676	0.02503	0.5600	2.2092	7.0453
4	129.7	0.2390	2.2964	0.02504	0.5679	3.3371	10.3467
....							
28	81.7	0.2503	3.3602	0.02534	0.8601	36.6425	49.5281
29	79.7	0.2508	3.4295	0.02535	0.8791	38.3818	49.6651
30	77.7	0.2513	3.5020	0.02536	0.8989	40.1598	49.6970
31	75.7	0.2517	3.5781	0.02538	0.9197	41.9785	49.6258
32	73.7	0.2522	3.6579	0.02539	0.9415	43.8397	49.4532

Maximum G^2/9266 happens at step 30th. This is the maximum mass flux at PRV orifice. It can be used to calculate the PRV orifice size as follows.

Gmax^2 = 49.697*9266=460492.5, Gmax =678.6 lb/sec-ft2
From calculated mixture specific volume (vi), it shows the relief is close to vapor. Let Kd=0.975.
PRV orifice (A, inch^2) = 42083/(25*0.975*678.6)=2.544 [Eq. (11)]

Required orifice area is about 2.54 in^2. L orifice PRV should be adequate for the relief service.

Note: Besides two-phase relief, this algorithm can also be applied to subcooled liquid, saturated liquid, or vapor relief.

Example 5 Rate an existing PRD for a new service.

Based on new relief rate and physical properties, we will calculate the required orifice area. If the required orifice area is less the existing one. The relief valve will be adequate for new service, provided inlet/outlet flange ratings are also adequate. Otherwise, either a larger PRD is required or an another PRD should be added. For PRV, we also have to check its inlet and outlet line hydraulic to make sure they are adequate for the service.

Answers to exercises of each chapter:

For Chapter 1:
1. See section I. 2. See section I. 3. No. Some projects only carried out to phase 1, 2, or 3, and stopped or cancelled for various reasons. 4. Phase 1 to 3, sometimes phase 5. 5. See section II and III. 6. See section II.2.1.
7. See section II.2.2. 8. See section II.2.3. 9. See section II.2.4 and III.2. 10. See section III.3. 11. See section III.4. 12. See section III.4. 13. See section III.5. 14. See section III.6. 15. See section III.7.

For Chapter 2:
1. See section I. 2. See section II. 3. See section III, III.1, and III.2. 4. See section III.1.2. 5. See section III.1.2
(1). 6. See section III.1.3. 7. See section III.1.4. 8. See section III.1.5 and IV.7.2. 9. See section III.1.8. 10. See section III.1.11. 11. See section III.2. 12. See section III.2.1.1. 13. See section III.2.2.1. 14. See section III.2.1.2. 15. See section IV.1. 16. See section IV.2. 17. See section IV.3 for centrifugal pump, and IV.7.2 for reciprocating pump. 18. See section IV.6.1. 19. See section IV.6.3 and III.1.7. 20. See section IV.6.4. 21. See section IV.7.2. 22. See section V. (2).

For Chapter 3:
1. See section I. 2. See section II. 3. For type and principle of compression: See section III.1a. For reasons they are selected: See section III.2 for centrifugal compressor, III.3 for axial compressor, III.4 for reciprocating compressor, III.5 for various rotary compressors. Ejectors are used for their low cost and abundant motive force fluid available. 4. See section III.1b. 5. See section III.2.1(1) for single stage centrifugal compressor, and section III.2.1(2) for multistage centrifugal compressor. 6. See section III.2.3, Fig. 6. 7. See section III.2.3. Many times a spill back line is provided to allow a centrifugal compressor operating away from its surge line. 8. See section III.3. 9. See section III.1a. 10. See section III.4.1. 11. See section III.4. 12. See section III.5. Major components are rotating elements, shaft, and casing. 13. See section III.5. 14. See section III.4.10. 15. Usually, the limitation is 5 to 10.

For Chapter 4:
1. See section I. 2. See section II. 3. See section III.1.1. 4. See section III.1.2. 5. See section III.1.4, Eq. (1-3). 6. TEMA. See section III.1.6. 7. See section III.1.4, Eq. (4a-b). 8. See section III.1.4. Basically, anything, which will increase fluid velocity, will increase overall heat transfer coefficient. 9. See section III.1.4. Basically, anything, which will decrease fluid velocity, will decrease pressure drop in a heat exchanger. 10. See section III.1.7. 11. See section III.2, III.2.1, III.2.2, and III.2.3. 12. See section III.2.1. 13. See section III.2.2.1-4. 14. See section III.2.3.1.

For Chapter 5:
1. See section I. 2. See section II. 3. Vessel shell, nozzles, and internals. 4. See section III.6. 5. See section III.7, Eq. (1a-c). 6. See section III.8, Eq. (1d-f). 7. See section IV.1. 8. See section IV.2. 9. See section IV.2, using mist eliminator (section IV.2.2). 10. See section IV.2.2. Three. Two. 11. See section IV.2.2, Eq. (5b). 12. See section IV.3. 13. See section IV.3.2, using coalescer pad. 14. It uses less area. 15. It is more efficient for vapor and liquid separation.

For Chapter 6:
1. See section I. 2. Single phase vapor line, single phase liquid line, two phase (vapor and liquid) line, and gravity flow line. 3. See section II. 4. See section III.1.1. 5. Eq. (3a) for liquid line, and Eq. (3b) for vapor line. 6. Eq. (6) for laminar flow, and Eq. (7a) for turbulent flow. 7. See Figure 2. 8. See section III.2.1. 9. See section III. 2.2.2. 10. Static head difference. 11. True. 12. Eq. (17). 13. See section III.4. 14. See section III.5. 15. See section III.6.

For Chapter 7:
1. See section I. 2. See section II. 3. See section III.1. 4. See section III.2. 5. See section IV. 6. See section IV. 7. See section V.1, Eq. (1-4). 8. See section V.1, Figure 8. 9. See section V.1. 10. See section V.1, Eq. (5-6). 11. See section V.1, Eq. (7). 12. See section V.1, Eq. (9a). 13. See section V.2, Eq. (13-16). 14. See section V.2, Y<0.667. 15. See section V.3, Eq. (18a-d). 16. See section VI.1 2) to 3). 17. See section VI.6. 18. See section VI.7.

For Chapter 8:
1. See section I. 2. See section II.1.1, II.2, II.3, and II.4. 3. See section I. 4. See section II.1.1, II.2, II.3, and II.4. 5. See section II.1.1.1 and II.1.1.2. 6. See section IX, item 1-2. 7. See section IX, item 3-5. 8. See section II.3. 9. See section II.4. 10. See section III.1.1. 11. See section III.1.2. 12. See section IV. 13. See section V. 14. See section VI. 15. See section VII.1. 16. See section VII.2. 17. See section VII.3. 18. See section VIII.2. 19. See section VIII.3. 20. See section IX.

1st revision list (Jan. 2013) for Process Design for Chemical Engineers (published July, 2012):

Ch. 2: p. 16 For II. (3): add ', or pump suction and discharge pressure.'
p. 16 For II. (10); add 'required pump hydraulic power'
p. 18 For (1) Change 'Impeller' to 'Impeller and shaft'
p. 23 For III.1.5 3rd paragraph, revise last sentence as follows: 'Cavitation usually is a problem for pumping saturated liquid, liquid with dissolved gases, or high pumping rate (>15,000 gpm). [1]'; last paragraph: add 'For conservative reason, many times NPSHA is calculated based on vessel bottom tangent line for vertical vessel or vessel bottom elevation for horizontal vessel. Consult a pump specialist or vendor for the required safety margin.'; delete 'See section IV.6.5 for more discussion.'
p. 26 For III.1.12 1st paragraph: change 'mathematical solution' to 'empirical equations'; at end of paragraph, add 'See section IV.6.5 for pump performance correction for viscous fluid.'; delete 2nd paragraph and add following paragraph: 'For viscosity less than 2000 SSU (440 centistokes), it is safe to use the NPSHR from the pump curve based on water. [1] In general, high liquid viscosity will increase pump NPSHR. ANSI/HI standard 9.6.7-2010 has more discussion in this area and has an equation can be used to calculate the NPSHR correction due to high viscosity. [18] Interested reader should read this standard or consult a pump vendor/specialist for advice.'
p. 28 last paragraph: change 'reduce the NPSHR' to 'increase its NPSHA'
p. 37 Delete section IV.6.5 NPSH Margin, add following paragraph:
IV.6.5 Pump performance correction for viscous fluid

From ANSI/HI standard 9.6.7-2010, Effects of Liquid Viscosity on Rotodynamic (Centrifugal and Vertical) Pump Performance, following equations are presented with Hydraulic Institute's permission [18].

These equations are used to estimate the reduced pump head, flow rate, and efficiency, and increased pumping power, when pumping fluid's viscosity is high (>32 centistokes (cts)). These equations are derived based on limited data for single-stage and multistage centrifugal pumps with closed or semi-open impellers and for following range of parameters: kinematic viscosity 1 to 3000 cst, BEP water pumping rate 13 to 18000 gpm, BEP water head per stage 20 to 430 ft. Testing fluids are Newtonian liquids, gels, slurries, paper stock, and other non-Newtonian liquids. They are empirical equations, but they give better estimation than the old HI method. The standard deviation for pump head correction factor (C_H) is 0.1, and for pump power requirement is 0.15 hp. Apply these equations to pump with specific speed (Ns) $\leq$ 3000, and assume a pump performance curves based on water are available.

Step 1. Calculate parameter B (pump performance Reynolds number adjusted for specific speed) as follow:
For multistage pump, use head per stage for H_{W-BEP}.

$$B = 16.5 \; (\mu/\text{s.g.})^{0.5} \; (H_{\text{W-BEP}})^{0.0625} / [(Q_{\text{W-BEP}})^{0.375} (N)^{0.25}] \qquad (12a)$$

If B≥40, the correction equations presented below are not applicable. A detailed loss analysis may be required. Interested reader should read ANSI/HI standard 9.6.7-2010 for this method. If B≤1.0, set H=Hw, Q=Qw, and E=Ew and go to step 5. [subscript 'w' is for water.] Otherwise, do following calculations.

Step 2 Calculate pumping rate correction factor (C_Q) and viscous fluid pumping rate (Q, gpm) as follows:
$$C_Q = Q/Qw = (2.71)^{[-0.165(\log_{10} B)^{3.15}]} \qquad (12b)$$
$$Q = C_Q * Qw \qquad (12c)$$
These two equations are valid for all water pumping rates (Qw) on the pump curves.

Analyzing test data, it is found C_Q equals C_H for viscous fluid at BEP. Therefore, following equations are valid.
$$C_{\text{H-BEP}} = C_Q \qquad (12d)$$
$$H_{\text{BEP}} = C_{\text{H-BEP}} * H_{\text{W-BEP}} \qquad (12e)$$

Step 3. Calculate pump head correction factor (C_H) using Eq. (12f) and the corresponding viscous fluid head at water pumping rate (Qw) greater or less than the BEP using Eq. (12g).
$$C_H = 1 - (1 - C_{\text{H-BEP}}) (Qw / Q_{\text{W-BEP}})^{0.75} \qquad (12f)$$
$$H = C_H * Hw \qquad (12g)$$

Step 4. Calculate pump efficiency correction factor (C_E) using Eq. (12h) and the corresponding viscous fluid pump efficiency (E) using Eq. (12i).
$$C_E = B^{(-0.0547 B^{0.69})} \qquad (12h)$$
$$E = C_E * Ew \qquad (12i)$$

Step 5 Calculate pump shaft input power (BHP) per Eq. (12j). This equation is valid for all pumping rates.
$$BHP = Q * H * \text{s.g.} /(3960 E) \qquad (12j)$$

These estimations should be considered as approximations. For more accurate estimation, consult a pump specialist or vendor. Interested reader should read ANSI/HI 9.6.7-2010 for more information.

p. 38 For Eq. (16): move '- Ha' to the end of the equation.
Ch. 3: p. 62 paragraph after Eq. (2c): Move 'k is isentropic exponent' to the end, and add 'at average suction and discharge conditions.'

Ch. 5: p. 92 For II: add '(3) Vessel nozzles: Provide a list of vessel nozzles and following information for each nozzle: number of nozzle required, nozzle size in inch, nozzle flange rating and type of facing, and a description of its service.' ; Move (7) into (10), renumber item (3) to (6) to item (4) to (7).
p. 102 For Figure 8b description: change 'vertical' to 'horizontal'.
p. 103 For IV.3 2nd, 3rd, and 4th paragraph: change 'coalesce pad' or 'coalescer' to 'coalescer pad'; change 'gravity settling method' to 'surge volume method'
p.103 to 105 delete IV.3.1 and add the following:
IV.3.1 Sizing liquid/liquid separator by surge volume method

From experience, required surge time for different liquid/liquid separation is listed in Table 1. From required surge time, the required total volume of the separator can be calculated using following equation.

$$VT = QF * tr / 60 \qquad (6a)$$
$$QF = QL + QH \qquad (6b)$$

VT is the required total separator volume, in ft3; QL is the light liquid design volumetric flow rate, in ft3/hr, QH is the heavy liquid design volumetric flow rate, in ft3/hr.

For a selected vessel length (L, in feet), vessel cross sectional area (AT, in ft2) and vessel diameter (D, in feet) can be calculated by following equations:

$$AT = VT / L \qquad (7)$$

$$D = 2 * (AT / PI)^{0.5} \qquad (8)$$

Prepare a table with a set of selected vessel length (L), calculated vessel diameter (D), and L/D ratio. Select a vessel its L/D ratio is from 3 to 5 as a preliminary size for the liquid/liquid separator.

A typical vertical liquid/liquid separator layout is shown in Figure 9a, and a typical horizontal liquid/liquid separator layout is shown in Figure 9b.

p. 105 For the title of IV.3.2: change 'coalesce' to 'coalescer pad'; last paragraph: delete 'a/b', change 'coalescer' to 'coalescer pad'; For Figure 9a/b description: change 'gravity settling method' to 'surge volume method'
p. 106 2nd paragraph (from section IV.4): change 'gravity settling method' to 'surge volume method', change 'coalescer' to 'coalescer pad'

Ch. 6: p. 115 After Eq. (16o), add following paragraph: 'Once fm is calculated, two phase flow line pressure drop can be calculated using Eq. (2), using fm (from Eq. (16l)), ρm_{cs} (from Eq. (16a)), and vm (no slip two phase fluid velocity, ft/sec, Eq. (16g)).'; after Eq. (16s), add: '[P1, P2 in psia, as Eq. 4d)]'
p. 125 For P1, P2, and Pavg, change psig to psia.

Ch. 8: p. 142 At 1st paragraph 3rd line and 2nd paragraph 1st line, add 'spring-loaded and pilot-operated' before 'pressure relief valve (PRV)'.
p. 155 For paragraph after Eq. (7b): after 'kc…at PRV inlet' add ', set kc to 1.0 for rupture disk sizing only'; For Eq. (8a): add ')' before '^0.5'.
p. 156 1st paragraph: after 'kc…at PRV inlet' add ', set kc to 1.0 for rupture disk sizing only'.

2nd revision list (Oct. 2014) for Process Design for Chemical Engineers:

Note: The first page number is from 1st revision book. Page number inside parenthesis is from 2nd revision book.
Ch. 1: p. 14 (p. 12) III.5 HAZOP analysis (was 'Hazardous analysis')
HAZOP stands for hazard and operability. HAZOP analysis is used to identify major hazard or operability issue related to current design and how to minimize hazard or operability issue to an acceptable level. ….

Ch. 2: p. 25 (p. 23) For III.1.8, 3rd paragraph:
There are two minimum pumping rates for a pump. One is due to the limitation of pumping liquid temperature rise and the other is due to pump hydraulic stable operation. Minimum pumping rate due to pumping liquid temperature rise (for pump discharge blockage and no cooling at pump return line) can be calculated using Eq. (5). Make sure to keep pump discharge temperature below pumping liquid bubble point temperature. ….
p. 26 (p. 24) For III.1.13 Sparing/priming (was Sparing), add 2nd paragraph:
Centrifugal pump is not self-priming. To start the pump, its casing and associated piping (especially the inlet piping) needs to be filled up with pumping liquid, and any trapped air/gas needs to be vented.
p. 35(p. 33) For IV.4, paragraph after Eq. (5):
During normal operation, pumping liquid temperature rise in pump is estimated using Eq. (5). Usually, this temperature rise is small and will not cause pump cavitation. But if pump outlet line is blocked and pump is still pumping, this will cause pumping fluid temperature rises above its bubble point and pump cavitation. To avoid this problem, a pump minimum flow (or return) line should be added at pump discharge line upstream the block valve to allow pumping liquid to return to its source. Let this return line flow through a cooler, if possible. Flow through this line should be sized to keep pumping liquid discharge temperature below its bubble point or for stable pump operation, whichever is larger. Usually, minimum pumping rate for pump stable operation is more critical. This minimum pump flow line can also be used to startup this pump. (revised)
For small pump or less expensive pump, pump minimum flow line is not justified. Plant can use pump operating procedure to prevent pump discharge block valve closed by accident; or to shut down the pump, if pump flow is below its stable pumping rate. For expensive or critical pump, pump minimum flow line is

usually provided. (added)

p. 37 (p. 35) For IV.6.5: 2nd paragraph, 4th line: ....1800 gpm (was 18000 gpm), ...

Ch. 3: p. 70 (p. 66) For III.1.4, Eq. (2) and Eq. (3):

$$Q = Wh * \int Cph * dT \qquad (2) \text{ (was } Q = Wh * Cph * (T1-T2))$$
$$Q = Wc * \int Cpc * dt \qquad (3) \text{ (was } Q = Wc * Cpc * (t2-t1))$$

.... $\int$ is integration. Eq. (2) is integrated from T2 to T1, hot stream outlet to inlet temperature, in °F. Eq. (3) is integrated from t1 to t2, cold stream inlet to outlet temperature, in °F.

Ch. 4: p.81 (p.77) For III.2.2.4, 5. Baffles:, paragraph after Fig. 7:

.... Smaller baffle spacing will cause more (was less) cross flow (more (was less) heat transfer) and higher shell side pressure drop. ....

Ch. 5: p. 95 (p. 90) For IV.1, last paragraph, 3rd line:

.... From this table, select a surge drum with the smallest vessel weight or L/D from 3 to 5. (add 'or L/D from 3 to 5')

p. 97 (p.9) For IV.2.1b, before Eq. (4f): Ignore the less sign (<) in Eq. (4e), ... (was 'From calculated vs,')

p. 98 (p. 93) For IV.2.1b, last paragraph:

.... From this table, select a separator with the smallest vessel weight or L/D from 3 to 5. (add 'or L/D from 3 to 5')

Ch. 6: p. 125 (p. 119) V., first few sentences: Line list provides a list of lines and line information in a unit. It is usually initiated by piping engineers. Process engineers will fill in following process information for each line: ....

Ch. 7: p. 139 (p. 13) For Eq. (18c/d): dpix = ... (was dpi = ...)

Ch. 8: p. 158 (p.152) For VIII., add a second paragraph:

The PRV inlet and outlet piping sizing criteria in section VIII.2/3 are for spring-loaded PRV. For pilot-operated PRV, consult manufacturer for its inlet and outlet piping sizing. In general, PRV inlet and outlet piping sized by VIII.2/3 may be conservative for pilot-operated PRV, since pilot-operated PRV has less blowdown and is insensitive to backpressure.

www.ingramcontent.com/pod-product-compliance
Lightning Source LLC
Chambersburg PA
CBHW081242170526
45165CB00009B/3161